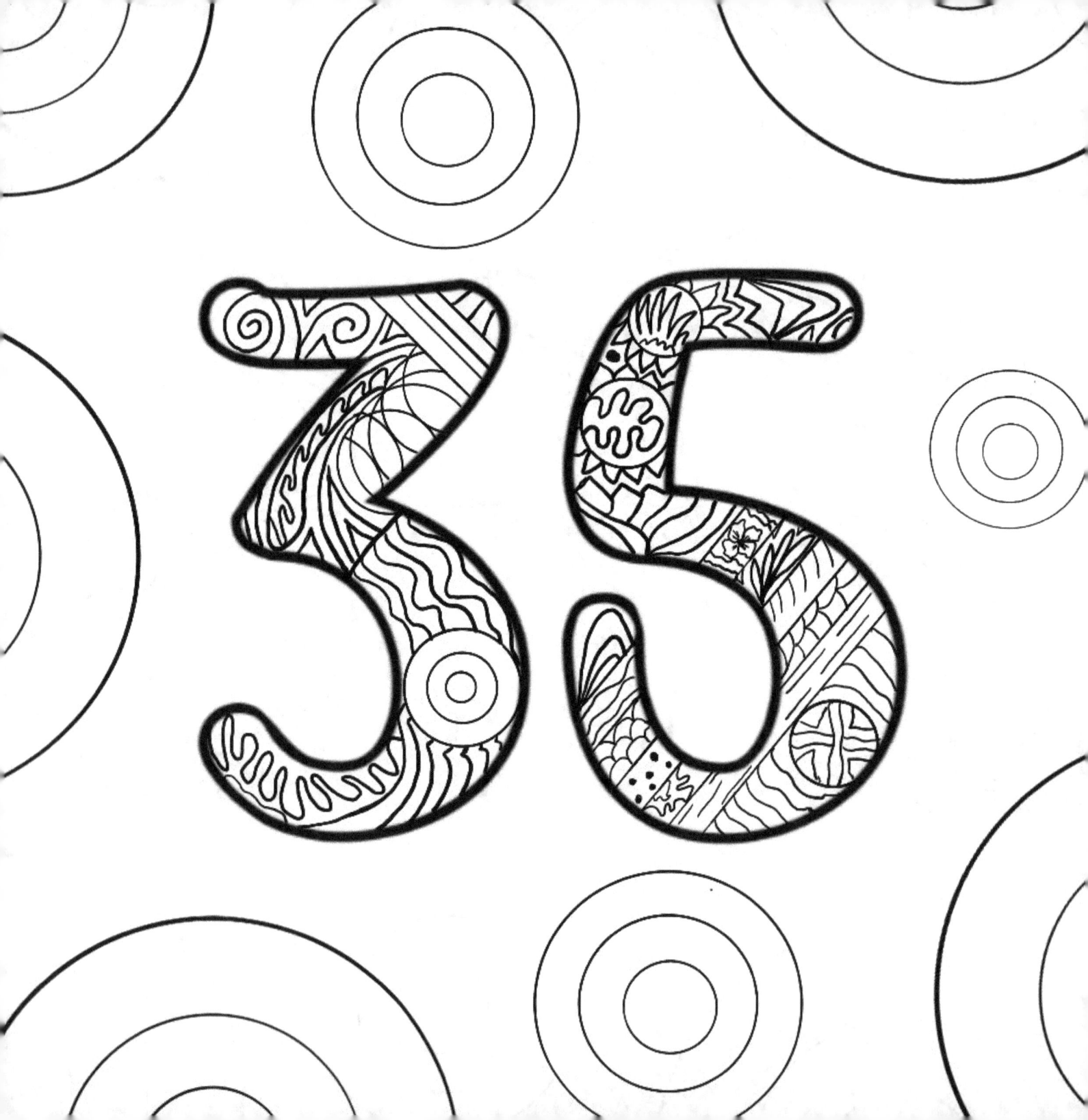

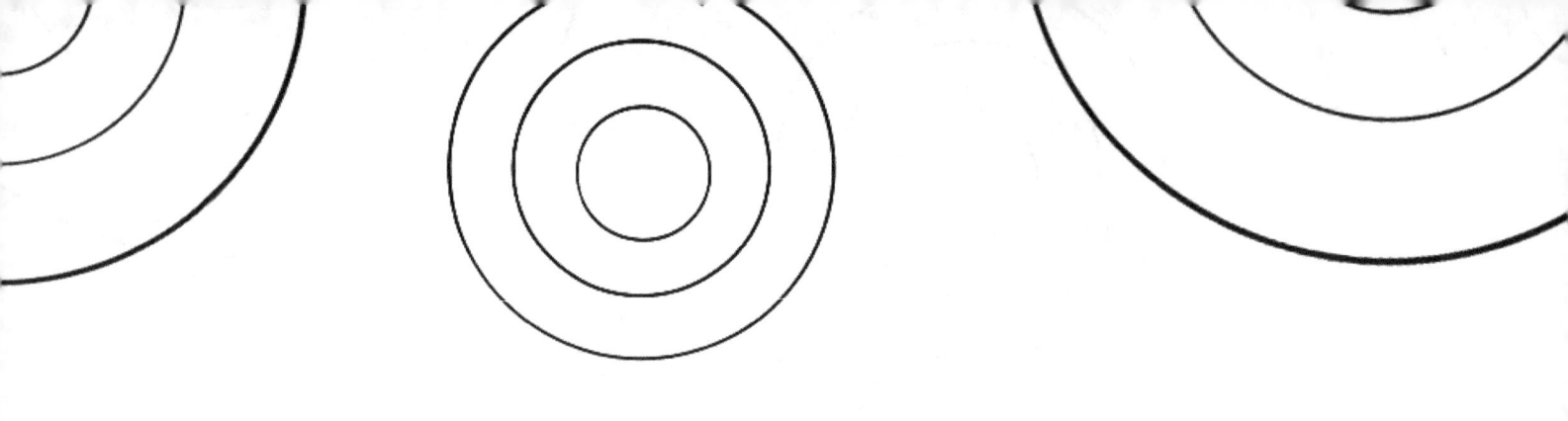

For Taarun xxx

"Mathematics is the language in which God has written the universe" Galileo Galilei.

I would like to acknowledge the Traditional Custodians of the continent of Australia. I recognise their continuing connection to the land, waters and culture and I pay my respects to their Elders past, present and emerging.

First Published in 2021.
Copyright © 1000 Tales

All rights reserved.
This book or any portion thereof may not be reproduced, stored in or introduced in a retrieval system, or transmitted, in any form or by any means without the express written permission of the illustrator.

ISBN: 978-0-6455554-9-3

www.ingramcontent.com/pod-product-compliance
Lightning Source LLC
Chambersburg PA
CBHW080847020526
44107CB00079B/2640